ROYAL
OBSERVATORY
GREENWICH

Aurorae

Bryony Lanigan

Royal Observatory Greenwich
Illuminates

First published in 2022 by Royal Museums Greenwich, Park Row, Greenwich, London, SE10 9NF

ISBN: 978-1-906367-96-1

At the heart of the UNESCO World Heritage Site of Maritime Greenwich are the four world-class attractions of Royal Museums Greenwich – the National Maritime Museum, the Royal Observatory, the Queen's House and *Cutty Sark*.

rmg.co.uk

Typesetting by ePub KNOWHOW
Cover design by Ocky Murray
Diagrams by Dave Saunders
Printed and bound in Spain by Grafo

About the Author

Bryony Lanigan is a science communicator with a love for everything in the Universe, from the very big to the very small. An astronomer at Royal Observatory Greenwich, she's also currently undertaking her PhD in atomic interferometry at Imperial College London. She has worked variously as a physics tutor, gymnastics coach, course coordinator and even glitter artist at music festivals, but always found a way to weave in her passion for physics.

About Royal Observatory Greenwich

The historic Royal Observatory has stood atop Greenwich Hill since 1675 and documents over 800 years of astronomical observation and timekeeping. It is truly the home of space and time, with the world-famous Greenwich Meridian Line, awe-inspiring astronomy and the Peter Harrison Planetarium. The Royal Observatory is the perfect place to explore the Universe with the help of our very own team of astronomers. Find out more about the site, book a planetarium show, or join one of our workshops or courses online at rmg.co.uk.

Contents

Introduction

Many people have 'see the Northern Lights' as something on their bucket list. These dancing, shimmering lights in the sky, known as the aurorae, have been the focus of human fascination for millennia and play a part in the spiritual beliefs of some of the oldest cultures in the world.

Part of their enduring mystery is a result of the difficulty in seeing them. The aurorae can only be seen in specific regions of the world, at very high latitudes. There aren't many people who live year-round in the harsh and unforgiving polar climes, and so, for centuries, many people only knew of the aurorae through tales told by explorers, travellers and traders, and what those individuals observed or learned from the communities native to those areas. While the modern world can feel small, the far north and far south of our planet are still incredibly remote, difficult and expensive to visit from anywhere else. Bright aurorae are only reliably visible above latitudes of 50°N and even then, that's only in North America – over Europe, you need to go even further towards the North Pole. It's even trickier in the Southern Hemisphere, with there being practically no inhabited land from which aurorae are almost always visible.

You'll have noticed that I said 'reliably', but that's perhaps a little misleading. The aurorae aren't visible every single day (or night!), but instead ebb and flow. Every day the **'auroral oval'** (I'll come to this a bit later) changes due to atmospheric conditions and it's these conditions that determine whether or not you can see the lights. They need to be just right to produce an auroral display.

From a scientific perspective, the aurorae still hold a few secrets. We have a pretty good understanding of what interactions generate the famous green, blue and even red colours of the lights, but the details are not simple and there remains much to learn about the process as a whole.

There is, however, a lot that we do know and to unpack it all we'll need to pick from a plethora of fields in physics and chemistry, starting with an exploration of **atoms** and energy and ending up far from our planet, extending our knowledge to other worlds.

Thousands of Years of Aurora Stories

Before we dive into the science behind aurorae, I want to briefly recount some of the stories people have told about the *Aurora Borealis* (the Northern Lights) and the *Aurora Australis* (the Southern Lights) for millennia.

It's worth recognising that these tales are not from long dead civilisations, but from ones still alive today. The Inuit of northern Canada, Alaska and Greenland (all regions where aurorae are common) are living, breathing peoples. Similarly, Indigenous Peoples of Australia, who have been on the continent for at least 65,000 years, have the longest continual cultural practices in the entire world. Their ancient stories about aurorae are supplemented by modern science, not superseded by it, and it would be a great disservice to these communities and their knowledge to claim otherwise.

Northern Hemisphere: Inuit

The Inuit occupy land spread throughout much of the northern part of North America and Greenland. There are multiple distinct Inuit communities, each with their own cultural practices and beliefs and therefore different aurora stories. It probably won't surprise you to know that when Europeans first came into contact with Inuit communities they made some incredibly racist assumptions. While there are many that deserve correction, the main one relevant to our story is the claim that Inuit are not interested in the sky. The opposite is true: there are, in fact, many stories that demonstrate the importance placed by Inuit on the stars and other celestial objects and phenomena.

Many native Inuit of Canada live inside the region known as the auroral oval, where the aurorae are brightest and most common. That far north, the day/night cycle that much of the rest of the world takes for granted is absent for much of the year. At high latitudes, more than 68° north of the equator, there are weeks during the summer with no darkness, and weeks during winter with no true sunlight. The former is known as the 'midnight Sun' and the latter as the 'polar night'.[1] During the polar night, the *Aurora Borealis* can be the brightest thing in the sky other than

---

[1] It is worth noting, however, that while there's no true sunlight for long periods, the sky is not pitch black at all times. There are stretches of twilight during the 'daytime hours'.

the Moon, so it's not surprising that they are a fixture in many Inuit mythologies. Even at the lower latitude of 60°N there are fewer than seven hours of daylight per day at this time of year.

The *Aurora Borealis* is sometimes described as almost otherworldly in its appearance and this sentiment is seen in several Inuit tales. The name given to aurorae in Inuktitut, the language used by Inuit in many parts of Canada, is *aqsarniit* or 'football players', which is a reference to the belief of these communities that the aurorae are spirits in the sky – specifically, spirits playing with a football made from a walrus skull. To some Inuit living in parts of Greenland these spirits, who they call *arsarnerit*, are exclusively children, while others believe that people of all ages may end up as part of the aurorae, often if they die in a specific way – perhaps violently, through childbirth, or in some other tragedy. While to most the aurorae aren't inherently dangerous or a bad omen if seen in the sky, some Inuit believe they have the capacity for harm if they get too close.

The aurorae aren't just a beautiful sight for the eyes. When it comes to the Northern Lights in particular, many people (Inuit to European, 18th-century adventurer to modern-day photographer) report hearing strange popping, crackling, or swishing sounds emanating from them when they are especially bright, particularly during calm and quiet nights. For a long time, scientists claimed that this was not possible, that it was simply the spooky ambience of being on the tundra late at night playing tricks on the brain...

Traditional explanations for the sound vary from place to place. Danish explorer Knud Rasmussen (1879–1933) spent some time in Iglulik (situated on an island of the same name off the northern coast of Canada) and, according to his accounts, the Iglulingmiut (inhabitants of Iglulik) tradition describes the sounds as being made by 'souls as they run across the frost-hardened snow of the heavens'. Stories from other communities say the noises are the spirits of the dead attempting to communicate with the living, whispering to those still alive. The link between sound and the aurorae is even stronger in other legends. Many Inuit traditions claim that you can call them close to you by whistling or scare them off by making clicking sounds or rubbing your fingernails together. Canadian Inuit often warn against tempting the *aqsarniit* down, lest an individual be taken up into the sky with them.

In recent years, it has finally been recognised that the sounds of the aurorae are more than just superstition or legend. Particularly strong aurorae can occasionally discharge, like the snap of a strong static electric shock, creating the eerie noises that are well known to those who see the aurorae frequently, confirming what Inuit and Arctic visitors have known for centuries.

Due to systematic efforts from colonists from the 16th until as late as the 20th century to wipe out the native population of Canada, we are lucky to have any record of these stories and knowledge. Inuit history, stories and cultural practices have historically been passed orally from one generation to the next and so the loss of individual people and the suppression of their language to the point of near extinction is literally the loss of their history and culture as a whole. There are many community organisations and institutions, including the Igloolik Oral History project in Nunavut (whose work I consulted), acting with urgency to record, preserve and revitalise centuries-old stories and cultural practices for future generations. Without such conservation projects, we'd be at risk of losing thousands of years of knowledge and learning.

Southern Hemisphere: Indigenous Australians

A moment ago, I noted that there wasn't anywhere in the Southern Hemisphere from where you could 'reliably' and 'regularly' see bright auroral displays. You may have wondered why I was so specific with my choice of words. Well, it's because if the conditions are right, you can see them from southern parts of Australia! While aurorae do not feature as heavily in the spirituality of Indigenous Australians, they are still present.

There are many different Indigenous nations across Australia with different traditions and spiritual beliefs about the aurorae. The brightness and presence of aurorae depends heavily on latitude, so where in Australia these nations lived affected the intensity and frequency of aurorae they saw, which in turn greatly influenced their interpretations of them.[2] Unlike in the Northern Hemisphere, there aren't any inhabited landmasses that lie within the auroral oval. As such, there aren't any peoples in the Southern Hemisphere who have the same familiarity with the *Aurora Australis* as Inuit of the Northern Hemisphere (who do live within the auroral oval) have with the *Aurora Borealis*. While it is still possible to see aurorae occasionally, bright aurorae, with intense colours and rippling curtains of light, are rare events.

---

[2] And Australia spans a lot of latitude. The northernmost tip of Australia is at a latitude of around 10°S, while the southernmost tip is a little over 43°S. Compare that with the UK: you can get a little higher than 58°N at the very northern part of Scotland and the southernmost tip of Cornwall dips just below 50°N.

Considering this, it makes sense that there are not as many cultural stories associated with aurorae from Indigenous Nations of Australia as there are from Inuit. The systematic killing of Indigenous peoples by British colonists that occurred even into the 20th century, as well as the deaths caused by diseases introduced to these communities, also had an impact and mean that most of the following come from accounts made by colonists rather than the originators of the stories themselves. Despite this, thankfully, the people and their traditions have survived and are undertaking massive efforts to document and revitalise their culture (as well as seeking recognition of the atrocities that were committed and what was lost). As part of this effort, ethnoastronomers, such as Duane Hamacher of the University of Melbourne, have documented many stories, helping preserve and maintain the spiritual beliefs of Indigenous nations.

In many historical and contemporary accounts, the appearance of the aurorae was and is seen as evidence of something occurring in the spirit world. According to accounts from white historians in the early to mid-20th century, the Ngarrindjeri people of coastal South Australia saw them as campfires in the land of the dead – partially because of the red colour, but also because of where in the sky they were seen. The land of the dead was considered to be an island off the south coast, and so, with the aurorae occurring over the southern part of the sky, the two were associated.

Sometimes the aurorae were seen as messages from the spirit world. The Pitjantjatjara people of the Central Australian desert tell of the appearance of poisonous flames towards the south when some hunters cooked a sacred *kalaya* (emu) – a warning for the rest of the tribe. The Gunaikurnai people of present-day Victoria, on whose land I was born, associate bright aurorae with fire: some see them as bushfires in the spirit world and others as harbingers of bushfires sent from the spirit world by an angry, powerful deity. It was said that *Mungan Ngour*, a powerful sky god, would send down these flames as retribution for breaking cultural law.

Even when they weren't associated with fire, the appearance of the aurorae was not considered lucky. Accounts written by colonists from the late 1800s and early 1900s say Indigenous peoples from nations in central New South Wales saw the faint red skyglow aurorae as either representations of great battles in the spirit world or victims of massacres on Earth rising up to another realm.

While different nations have different teachings around aurorae, most tend to view their appearance as a bad omen. This is possibly because of the rarity of bright, visible aurorae in the region as well as the typical colour. The aurorae that are seen in northern parts of Canada are typically green in colour and often occur in beautiful arcs or rippling curtains across the sky, while the faint red glow of the Australian aurorae no doubt contributed to their distinct interpretation among its peoples. Red is not a lucky colour to many Indigenous Australian groups and so it's hardly surprising that seeing an occasional red glow in the sky would be taken as a sign of danger to come.

The spiritual beliefs and accounts that I have presented here are a tiny taster of the wide world of ethnography. I have included examples from two cultures, but there are many more with beliefs about the aurorae: the Sámi of Finland, the Aleut of the Aleutian Islands, the Yu'pik peoples of Alaska and Russia, to name just a few! While many would consider ethnography to be disparate from astronomy (and indeed many other branches of science) this isn't the case. Every culture has looked up to the sky and found inspiration and guidance from it, and, more and more, there is a push to recognise and document this – so, if you're interested in learning more, there is a wealth of information waiting for you.

For this book, though, we will leave the cultural astronomy there and move on to the technical side – let's start talking atoms.

Our Aurorae

The aurorae can be explained in a single sentence: solar wind from the Sun interacts with Earth's atmosphere to produce light. That's the aurorae in a nutshell. But while that sentence does technically describe what's going on, there's a lot to unpack.

What is this solar wind and how is it created? How does it interact with Earth's atmosphere and why does it produce the colours that it does? What do we mean by Earth's atmosphere anyway? Why do the aurorae only appear in certain regions – why not all around Earth?

To answer those questions, we will delve into **electromagnetism**, geophysics, some atomic physics and I'll throw in a little bit of nuclear fusion for good measure. Let's get started!

Fundamental Physics

Light

Before we can start to talk about the hows and whys of aurorae, we first need to ask a simple question: what exactly is light?

In everyday conversation, it's pretty obvious what we mean when we say light: we're talking about that thing that lets us see. To physicists, though, this is actually a very specific type of light, creatively named 'visible light', and it is just one part of the **electromagnetic (EM) spectrum** (see Figure 1). This spectrum also includes microwaves, radio waves, x-rays and more, the vast majority of which are invisible to the human eye. The divisions on the spectrum are made on the basis of energy. EM waves like radio waves, for example, have a much lower energy than ultraviolet light, which is in turn lower energy than x-rays. The important thing to note here is that different types of light are produced by different processes and have different energies.

Each section on the spectrum can be subdivided even further according to energy and for the visible part these different energies correspond to different colours! When we talk about light, we often speak of it not in terms of energy but in terms of **wavelength**. The wavelength of a wave is, in a manner of speaking, its 'size'. The shorter the wavelength, the higher the energy and vice-versa. Radio waves have relatively long wavelengths, typically a few metres or more, while ultra-high-energy gamma rays have

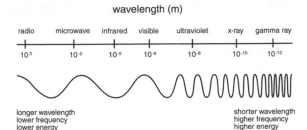

Figure 1: The electromagnetic spectrum.

wavelengths smaller than the size of an atom! When we talk about visible light, we're talking about wavelengths of a few hundred nanometres (a nanometre is equal to one billionth of a metre) – that's only a fraction of the width of a human hair.

But what exactly is the mechanism by which different things produce different energies and wavelengths of light?

To explain this, we need to get right down to the very base of what things are made of: atoms. Atoms are made up of a dense, positively charged centre (**nucleus**), composed of particles called **protons** (positive particles) and **neutrons** (neutral particles), with one or more **electrons** (negative particles) 'orbiting' it.[3]

Different elements have different numbers of protons in their nucleus, which determines their **atomic number**. Hydrogen, with an atomic number of one, has one proton inside its nucleus, helium has two, lithium three and so on. Rather than affecting the type of element the atom is, the neutrons found in the nucleus instead affect its stability. Atoms with the same number of protons but different numbers of neutrons are referred to as different **isotopes**. Luckily for us that particular terminological headache won't be referred to again in this book. Huzzah!

For an atom to have no electric charge, it must have equal numbers of positive

-----

[3] I've put orbit in inverted commas because the electrons aren't really orbiting the atom's nucleus, it's more that they are existing vaguely nearby… but that's the subject for a different book, one completely unrelated to astronomy!

and negative particles – equal numbers of protons and neutrons. This isn't to say that there is no such thing as a charged atom. These are known as **ions** and they contain different numbers of protons and electrons. Take the example of lithium (Li): an electrically neutral lithium atom has three protons and three electrons. You can, however, get lithium ions with four electrons, which are electrically negative, or lithium ions with two electrons, which are electrically positive.[4]

If there is an electron near an electrically positive ion (i.e. an ion that has fewer electrons than protons), then, generally speaking, that electron will want to join that ion to make it electrically neutral – opposites attract and all that. However, in very hot conditions (we're talking physics hot:

---

[4] You can keep on adding electrons and making the ion more and more negative. Likewise, you can continue taking them away until you've got a bare nucleus!

thousands of degrees Celsius, not British 'heatwave' hot), the electrons will have enough thermal energy to completely ignore this desire and make their own way in life. This means that in hot conditions, you have fewer atoms and more ions (or even bare nuclei), with electrons independently zipping around as they please.

To put it succinctly: different numbers of electrons = different ions, different numbers of protons = different atoms.

A basic diagram of an atom shows a nucleus at the centre with electrons in rings around it (Figure 2). While this arrangement is not fully correct, it is close enough and a good enough jumping-off point to discuss the motion of electrons within atoms.

Each of the rings around the nucleus corresponds to a very specific energy level. The electrons closer to the nucleus are at a lower energy level than those further away. One way of showing this is to draw a predictably named 'energy level diagram'

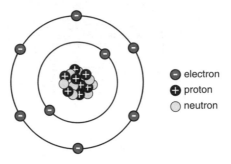

Figure 2: A basic diagram of an oxygen atom. Negatively charged electrons orbit the nucleus, which is composed of protons (positive charge) and neutrons (neutral).

(see Figure 3). The lowest energy level is at the bottom, with increasing distance from the nucleus correlating to increased energy levels.

Electrons have preferences for the state in which they exist and existing in states further away from the nucleus requires more energy.[5] Think of the different

---

[5] The electrons don't all crowd into the states closest to the nucleus, despite the fact that they seem, in theory, to be at 'the lowest energy'. Electrons are, on the whole, social-distancing pros that can't stand the idea of overcrowding and so obey pretty strict rules as to how many can be in each

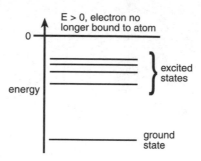

Figure 3: Energy levels of electrons inside an atom, from ground state to more excited states. More negative energy states are more tightly bound to the atom, while an electron with zero or higher energy is no longer bound to the atom.

states like a set of stairs, with each state represented by an individual step – if you want to climb further up the stairs, you need a certain amount of energy to climb each step. If electrons absorb these very specific amounts of energy, they can 'climb' the stairs. Similarly, when they 'descend' the stairs they lose specific

energy state, which override any desire to be in the absolute lowest energy state. Instead, we say they will be in the 'lowest available energy state'.

amounts of energy with each step.[6] Do note, though, that electrons don't need to rest on every step on their way up or down – they can skip one, two, five – as many as they wish! The important thing is that the specific amount of energy they absorb or lose is exactly the difference in energy between the two steps or states.

Now, saying 'specific amount' gets a bit confusing. What amount? At this point, I'd best introduce the **photon**, which you may or may not have encountered before. One way to think about light is to envisage it as a series of little bowling balls that bounce around, each carrying a set amount of energy. Atoms can absorb these little bowling balls to allow their electrons to move to higher energy levels, or emit them to reduce their energy. The miniature bowling balls emitted can then collide with the structures in our eyes,

---

[6] If you're interested in reading a little more about energy levels, take a look at p. 125.

allowing us to see! This metaphor is tortured enough already, so we'll stop it there, but there is very good reason for us to imagine light as these discrete packets that we call photons. By absorbing and emitting light, atoms gain and lose energy – light and energy are almost two sides of the same coin here.

You can't absorb half a photon – you either absorb it or you don't. If a photon hits an electron and it doesn't have the right amount of energy for the electron to transition to a different energy level, then the electron will just completely ignore it. If you only have enough energy to lift your foot halfway up the step, you won't be able to climb to the next level and you can't just stand between the two!

That's a strange thing to get your head around, but stick with me here: it's about to get even stranger.

Unlike your standard staircase, there are no health and safety regulations stating that there must be equal spacing between different energy levels and there most definitely is not. Depending on which levels the electron transitions between, different amounts of energy are required… or, in other words, different colours of light!

It's not as though every single element has the same internal electron energy levels, either. The energy required for electrons to move from the lowest state (called the **ground state**) to higher states (**excited states**) varies between elements. Of course, when you start to combine atoms together to form **molecules,** the energy levels change once more! This means that different elements will absorb or emit at different energies, or different colours of light.

Mathematically and experimentally, we can calculate and measure the **absorption** and **emission spectra** – the energies of light that are absorbed or emitted – of different things (atoms, molecules, etc.). These are like barcodes and they are unique to each substance or compound. So, when we observe light coming from something unknown, we can compare the 'barcode' to our known spectra and use it to identify whatever is emitting the light.

Electrons in excited states are not stable: given enough time, an electron will emit a photon and drop down to the lowest available energy level, a process called 'spontaneous emission'. Spontaneous emission isn't the only way to lose energy, though. If an electron collides with another electron, or even an atom or molecule, it can be forced to emit the extra energy. This also works the other way around – absorbing a photon isn't the only way for electrons in an atom or molecule to gain energy. If an atom or molecule in the ground state is hit by an electron, then its electrons can absorb some energy and be excited into higher energy levels. This is called **collisional excitation**.

As a quick summary: we can think of light as little particles called photons, with different energies that correspond to different colours of light. The fundamental building blocks of nature, atoms, are composed of smaller particles – protons, neutrons and electrons, with the protons and neutrons being confined to the central nucleus and the electrons 'orbiting' around them. Electrons orbit in different 'energy levels' and moving between them requires either the absorption or emission of photons of a specific energy.

Okay, so we've got the very basics of light and atoms, but there's a little more fundamental physics that we will need to discuss before we can even start to consider the aurorae. That little bit happens to be one of the first real physics concepts that you encounter in school: magnetism.

Magnetism

When you first came across magnets, they probably seemed like magic. 'You mean, I can pick up small things using this stick of metal?! And, if I turn it around, then I can push them away? What's more, if I have two of these magic sticks, then they'll be attracted to each other at one end but repel at the other?!' You probably spent an inordinate amount of time desperately trying to get the two ends that repel each other to touch and finding that it was almost impossible.

Then you learnt that it's not magic-magic, but that there is an invisible force acting on all these objects. You learnt that there are two poles, a north and a south, and that opposite poles attract while like poles repel.[7] To understand these

---

[7] Some people feel that learning the science of how magnets (and many other things) work 'ruins' them. While I admit that studying physics may have made me a bit more of a killjoy when it comes to bad science in film, I could write an entire book about magnetism and how fascinatingly bizarre it is. Here I will limit myself to the fundamentals.

processes of attraction and repulsion, we need to discuss this force in a little more detail. On a diagram of a single bar magnet, we draw magnetic field lines around it that flow from the north pole to the south pole (see Figure 4). Magnetic materials that come near the magnet will interact with these field lines.

Suppose we have a bar magnet sitting on a table and we bring a second bar magnet close to it such that the second's south pole is near the first's north pole. The second magnet will experience a force that will pull it even closer to the first's north pole thanks

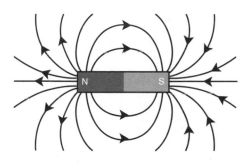

Figure 4: Bar magnet with magnetic field lines flowing from north to south.

to a change in the magnetic field lines. Rather than curving from the north to the south of their own magnet, the field lines will instead be drawn to the opposite, and closer, pole of the secondary magnet. This results in a force that pulls the magnets towards each other, so that they join to make one big bar magnet (see Figure 5).

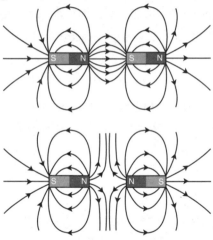

Figure 5: Two bar magnets interacting. When two opposite poles are brought together, the field lines change such that they are connected to the nearest opposite pole, producing an attractive force. When two like poles are brough together the field lines resist being squashed together and exert a force that keeps the magnets apart.

34

If instead you try to push both north or both south poles together, you will find that the magnets resist, opting instead to skew. If you've ever heard the phrase 'opposites attract', here is the perfect example! Opposite poles attract, while like poles repel. (The same goes for electric charges, too – opposite charges attract, while like charges repel.[8])

This is because the magnetic field lines are essentially getting in the way as you try to push two like poles together – there needs to be space for them to go from north to south and if two south poles are

---

[8] Why aren't negatively charged electrons pulled in towards the positively charged nucleus, before falling in, combining with the protons and destroying the atom in the process? This was a problem in early quantum theory, as it's pretty clear that atoms do exist. Nobel prize-winner Erwin Schrödinger managed to solve it by assuming that while there *is* a force that pulls the electrons towards the nucleus it is balanced by the electron having some angular momentum that counteracts it.

touching that isn't possible! The magnetic field lines can be squashed and bent around a bit, they can change what they connect to, but it's really, really hard to break them.

So far, we've been talking about the magnetic fields generated by something like a bar magnet, however this isn't the only way that we get magnetic fields. They are also generated by moving electric charges, otherwise known as electric currents.

This is an effect that can be observed in **electromagnets,** which can produce powerful magnetic fields just from the flow of electricity. They are used in a variety of medical and practical contexts, from MRI machines to supercomputers. The strength of the magnetic field generated depends on, among other things, the speed and strength of the current (by which we mean charged particles moving en masse). In practical terms, this means the individual or object operating the magnet has a level of control that simply does not exist with bar magnets. You can hardly 'turn off' a bar magnet, but by stopping the current you can turn off an electromagnet.

So electric currents generate magnetic fields, but the opposite is also true: moving magnetic fields can generate electric currents! This is where the physics gets a little bit loopy – if electric currents generate magnetic fields and magnetic fields generate electric currents, then won't these electric currents also generate magnetic fields?

The answer is yes, they can! What's more, they'll be induced in the opposite direction to the original! However, just because these things *can* be induced, it doesn't mean that they are particularly strong. These second-order inductions are much weaker than the original current, and so, while they can have an effect, that effect isn't enough to completely counteract the original current and can therefore be ignored in simple cases.[9] The most important result of this for us is that electrically charged particles are affected by magnetic fields and can effect change in those magnetic fields (if they are fast enough).

Until now, we have been looking at currents, the movement of charged particles en masse such as in a wire. However, it's also helpful to consider the case of a singular charged particle moving through a magnetic field.

The most straightforward example in physics is that of a charged particle moving through space in a straight line (see Figure 6). If that particle were to pass through a

---

[9] In many practical cases second-order inductions can't be ignored, which is why electrical engineers have spent so much time finding ways to mitigate the induced currents and magnetic fields, but we won't dwell on that too much here.

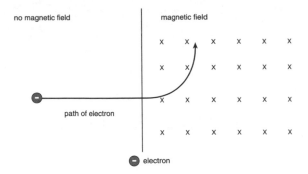

Figure 6: Charged particle moving through a magnetic field.

magnetic field, its path would be bent to a degree determined by its charge and the strength of the magnetic field that it is moving through. As well as altering the course of a particle, magnetic fields can also accelerate or decelerate them, which is most definitely important for auroral activity!

There is a heck of a lot more to it than what I've just told you, but I think we now have enough fundamental physics to start applying what we know to the aurorae.

Aurorae. Almost...

The Sun

Let's go back to our one-sentence definition of aurorae: solar wind from the Sun interacts with Earth's atmosphere to produce light. Now that we've covered some of the basics, let's tackle this sentence from the very beginning with the creation of the solar wind by our Sun.

Our Sun is a star[10] and as such produces its own light through a process

---

[10] A G2V yellow dwarf star to be exact, though for more information on that you should check out another book in this series, *The Sun*! Am I doing this marketing thing correctly?!

called **hydrogen fusion**. It is composed of hydrogen and helium gas at a ratio of roughly 3:1. At the very centre of our Sun, where temperatures are in excess of 15 million °C (very hot) and pressures in excess of 260 billion Pascals (very dense), the atoms do not exist as nice, well-defined, separate entities, but instead as a kind of plasma soup, with the nuclei separate from the electrons. This plasma is always in motion, never still.

Under normal circumstances, what we would expect is for the nuclei to be doing their level best to stay as far away from each other as possible. Positive charges are repelled by other positive charges and we would expect this electrostatic repulsion to ensure that they never get very near each other.

But these are not normal circumstances and nuclei forget their manners. Rather than turning in the other direction, thanks to the pressure and temperature at the centre of a star, there is a chance that they could fuse together – not exactly the polite, detached behaviour we have come to expect from them. Through a convoluted process, four individual hydrogen nuclei become one brand-spanking-new helium nucleus! Hydrogen fusion occurs pretty much exclusively at the centre of a star – it's simply not hot or dense enough anywhere else for hydrogen atoms to be forced together to form helium.

This brings us to a central concept in physics – whether an event can be described as 'rare' depends not only on how likely it is to happen to a specific object, but also on how many of those objects exist. If something happens to people at a rate of 1 in 1,000 – say, winning more than £100 from a lottery ticket – statistically speaking, this would mean that if 1,000 random people were gathered in a room one of them would have experienced it. Now suppose there are 10,000 people – again, statistically speaking, this would be true of ten of them, and so on and so forth.[11]

This is at the heart of many strange things in physics – something may only have a 0.00001% chance of occurring at any one time and therefore be quite 'rare' in isolated experiments, but in the real world, where millions of processes take place all at once, that 0.00001% suddenly really matters!

Returning to our discussion of the Sun, it's still pretty unlikely that any individual hydrogen nucleus will be forced to fuse with any other one, but there are so many of them that it ends up happening quite a lot. The fusion process involves several steps, but for our purposes we care mainly about the end products: a helium nucleus

---

[11] It's why the compliment 'You're one in a million!' isn't really all that complimentary. What you're really saying is that there are almost 8,000 people on this planet just like me!

and some energy. In the normal world, we think of creating something new as requiring energy, but hopefully by now you won't be too surprised to learn that the quantum world messes with that as well. A star like our Sun doesn't shine because it's burning anything, but because it is fusing hydrogen into helium at a rate of around 600 million tonnes every single second – that's what it 'runs on'.

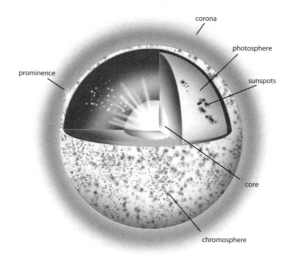

Figure 7: The structure of the Sun.

The energy created by fusion is made in the core, though, and in order for it to be emitted as light, it needs to make its way from the very centre of the Sun to the outermost layers. Because the light can't just travel in a straight line this is a process that can take anywhere from several thousand to several million years.[12]

Once the energy generated by fusion in the Sun's core has made its way to the outer layers, it reaches the 'surface'. Unlike Earth, the Sun doesn't have a proper surface, at least not a solid layer that you could stand on. What we can do instead is build up a 'surface' with three main layers: the **photosphere, chromosphere** and **corona** (from innermost to outermost, see Figure 7). The photosphere is what we 'see' when

---

[12] It's worth noting that, strictly speaking, by the time light leaves the Sun it won't actually be made up of the same photons that were created initially. This is because the photons are constantly absorbed and re-emitted. Only some eventually make their way out.

we look at what the Sun is emitting in the visible part of the spectrum, so it makes sense that it's called the *photo*sphere! (Confusingly, we frequently refer to the Sun's photosphere as both its surface and the first layer of the atmosphere.) Temperatures here reach around 4,000 to 6,500 **Kelvin** (K) (3,700 to 6,200 °C), with the heat being less intense in the photosphere's outer regions.

The next layer, the chromosphere, does emit some faint visible light, you can, however, only see it during a solar eclipse, when the main bulk of the Sun is blocked by the Moon and just the star's faint edge can be seen from Earth. Interestingly, while the temperature has been cooling as we make our way from the centre of the Sun to the edges, this trend doesn't continue through the chromosphere. Instead, temperatures in the chromosphere are higher than in the photosphere, and the corona, the outermost layer, is much hotter than the chromosphere and photosphere, reaching temperatures of hundreds of thousands of Kelvin!

Observations have shown that there is a very narrow (perhaps as narrow as 100 km) transition region between the chromosphere and the corona where the temperature skyrockets from around 8,000 K to something like 500,000 K, but exactly why the corona is so much hotter than the chromo- and photospheres is still not known. It's thought to be due to the extreme electric and magnetic conditions found in the Sun, but we don't currently have the physics to explain it and the search for theories that withstand the heat (pun intended) is ongoing.

Our Sun has its own magnetic field, but, because it is made of incredibly hot, charged, gas, its magnetic field is nowhere near as simple as that of a bar magnet. All of the streams of plasma moving across the Sun create electric currents, which in turn generate magnetic fields that loop off the photosphere and back again. Beyond this, streaming away from the corona, is the solar wind, a super-hot plasma made of charged particles that extends out into space forming the **interplanetary magnetic field (IMF)**. You can think of the IMF as a huge extension of the Sun's own magnetic field. In some ways, the IMF and solar wind are the same phenomenon – the latter carries the former out into space, with charged particles 'surfing' on its magnetic field lines. These charged particles are primarily protons and electrons, though there are some traces of heavier elements such as carbon, oxygen and even iron!

The solar wind travels at incredible speeds – hundreds of kilometres every second, far surpassing the speed of sound. While it is always being emitted, its strength varies. Our Sun goes through an 11-year cycle, with sunspot numbers, magnetic activity and solar winds peaking in the middle, a period known as the solar maximum. **Sunspots** are regions of the Sun that are temporarily cooler and therefore appear dimmer than the rest of the surface, but they are also known to be sites of intense magnetic-field activity. In fact, it is thought that the stronger magnetic fields around sunspots cause them to be cooler and therefore darker. They usually appear in pairs on opposite hemispheres of the Sun, with opposite magnetic polarity, one with a north pole and the other a south pole, and magnetic field lines connecting them. And so, the more sunspots visible on the Sun's surface at any one time, the more chaotic the magnetic terrain of the Sun will be.

Alongside sunspots, during the peak of the **solar cycle**, we see a lot more **coronal mass ejections** (**CMEs**) and **solar flares**.[13]

---

[13] While things like solar flares and coronal mass ejections are linked to sunspots and frequently occur alongside them, they can appear at other times too and we don't have a full understanding of how they work.

A solar flare describes the moment when a large amount of electromagnetic energy (light) is emitted suddenly from the Sun, while coronal mass ejections are when large amounts of plasma (hot gas) are emitted. CMEs have a greater effect than flares on the solar wind; the more plasma expelled from the Sun, the more particles in the solar wind. This is very important for the aurorae.

So we've got our extremely hot ball of superheated plasma, constantly spewing charged particles into space and at times coughing up a larger chunk. These particles, the solar wind, travel through interplanetary space to… Where exactly?

Earth

To us! The particles of the solar wind don't fly straight as an arrow – remember that these charged particles are affected by magnetic fields, so they travel along the interplanetary magnetic field, interacting with and changing it as they go. The IMF isn't the only magnetic field they will encounter, though – we have our own one here on Earth too.

Earth's magnetic field

At its most basic, Earth is just a giant bar magnet, with a north pole and a south pole. The Earth's magnetic north and south poles do not correspond to the geographic North and South Poles and are also changing in position. To add further to the confusion, what we call the 'North Pole' is technically the south pole of the magnet that is our Earth!

Earth's magnetic field is driven by an **internal dynamo**, its outer core. Composed of a mix of iron and nickel, the core is constantly in motion, with complex currents driven by the planet's rotation and heat exchange between

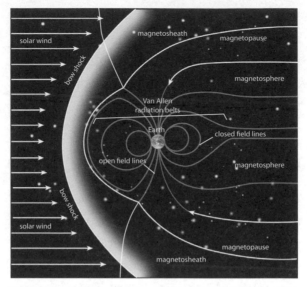

Figure 8: Earth's magnetosphere is shaped by the interaction of solar wind with the planet's magnetic field.

different layers. (We're talking here about convection currents, circling currents that bring cooler material up and hotter material down.) Thanks to the incredibly hot temperatures found there, electrons dissociate from the nuclei, resulting in a charged soup sloshing around, which generates the magnetic field.

Earth's magnetosphere

Now, that is the Earth's magnetic field, but what we really care about here is the **magnetosphere**. These are subtly different: Earth's magnetosphere is defined as the region of space around the Earth where the dominant magnetic field is that of Earth rather than that of the Sun.

The shape of the magnetosphere is a consequence of the interactions between Earth's magnetic field and the solar wind (see Figure 8). We can see that it is not symmetric: on the side facing away from the Sun the magnetic fields lines are elongated and stretch further from the Earth than those on the side facing the Sun. Its appearance is similar to what you see when running water comes upon a rock. Encountering an obstacle, the water flows to either side.

The boundary of the magnetosphere is known as the **magnetosheath** and is constantly shifting with the solar wind. On the daytime side of Earth, the side turned towards the Sun at any one time, it is typically anywhere from around six to ten Earth radii away from the planet's surface. Meanwhile, on the night-time side, it lies well beyond 50 Earth radii! When particles on the solar wind reach the magnetosheath most of them are slowed down and redirected to pass around us. This creates a region we refer to as the 'bow shock', where the speed of the solar wind, until this point supersonic, is substantially reduced.

Even as the solar wind is compressing Earth's magnetic field and, for the most part, being redirected either side of our planet, some little bits still manage to sneak through the magnetosheath. Much of this material then gets trapped in the magnetic field, forming the Van Allen radiation belts. Named after the space scientist, James Van Allen (1914–2006), credited with discovering them, these are regions around Earth that contain lots of energetic charged particles.

Taking a closer look at the magnetic field lines of the Earth, we can see that some, at first glance, seem to break the rule we set for ourselves earlier. Remember that magnetic field lines must always connect to a north and a south pole. These ones instead appear to stream off into nothing. What I've failed to mention is that this diagram is incomplete: we're only showing Earth and there are plenty more magnetic things out there for these 'loose' magnetic field lines to connect to.[14] The magnetic field lines that connect back to Earth

---

[14] In preparation for writing this section, I read a paper entitled 'Whose field line is it anyway?' (https://academic.oup.com/astrogeo/article/45/3/3.36/237191). I thought more people should be aware of that.

are 'closed', while those that stream off into space are 'open'. It is the boundary between the open and closed field lines that is crucial for the aurorae – this is where they are most likely to occur.

Earth's atmosphere

By this point in our story, the solar wind has made its way through space, all the way to Earth's magnetosphere. Part of it, the part we care about now, is being funnelled towards the poles. As it approaches Earth, it will encounter our atmosphere.

Earth's atmosphere is made up of a cocktail of different gases but, most importantly for both the formation of aurorae and the continued survival of humans, it is around 78% nitrogen and 21% oxygen. The remaining 1% includes traces of argon, carbon dioxide and water vapour. There aren't uniform amounts of these elements in all parts of the atmosphere, though. There's a lot more water vapour at lower altitudes than at higher ones, for instance, and the concentration of nitrogen also drops the further you get from Earth's surface.

Rather than existing as single atoms, many of these gases are in simple molecules. Nitrogen is almost always found in molecular form, as N_2 (two nitrogen atoms joined together). Down near Earth's surface, where we mere mortals live, oxygen is also usually found in molecular form, O_2, but higher up you can find atomic oxygen floating free. It is actually one of the most 'common' elements by weight when you are very high up! That doesn't mean there's all that much of it, but in the extremely thin atmosphere of high altitudes, there's very little of anything else!

We can split the atmosphere into various layers and define each by its characteristics and by the events that usually take place there.

The **troposphere** is the lowest layer of the atmosphere and is where most weather occurs. By weight, it accounts for 75% of the atmosphere, even though it's easily one of the thinnest sections – between 6 and 20 km deep depending on where you are on the planet. The top of the troposphere is called the **tropopause** and here temperatures reach as low as -80 °C!

Continuing upwards, we have the **stratosphere,** home to the ozone layer, high winds, and some bacteria.[15] As we move up through the stratosphere, something strange happens: the temperature increases with altitude. Despite being around 50 km

---

[15] Yes, you read that right: bacteria live high up in our atmosphere! It gives us yet another thing to look out for in our search for life on other planets – even if a planet's surface is hostile to life, we can look for bacteria in its atmosphere. There are currently researchers looking into that exact possibility for Venus!

from Earth's warm surface, at the boundary between the stratosphere and **mesophere** (the next layer) the temperature peaks at around -5 °C.[16]

The region we denote the mesosphere is defined by changes in temperature. It starts at the point where temperature stops increasing with altitude and instead begins to decrease and it stops at the coldest point of our atmosphere, known as the **mesopause**, where temperatures are around -143 °C. It's here, around 80 to 90 km above Earth's surface (or rather, just above this height), that the conditions are right for aurorae.

Beyond the mesosphere, we enter the **thermosphere**, where charged particles from the Sun and ultraviolet radiation

---

[16] The reasons for this are complex, to say the least, but it essentially boils down to the fact that our definitions of temperature are much more complex than just how hot something feels and are instead related to the speed of the various particles.

mean that temperatures can vary wildly depending on solar conditions – in places they are as high as thousands of degrees Celsius! By the time we get to the thermosphere, however, the atmosphere is incredibly thin. Because the particles are so spread out they cannot transfer heat as easily down through the rest of the atmosphere – lucky for us, or we'd be in dire straits!

A question I am frequently asked by curious eight-year-olds is 'Why doesn't the atmosphere float away from Earth?' It's a pretty good question! Gas is often defined as a material that expands to fill the container it is in, so why doesn't our atmosphere, being made of gas, just expand away from Earth in an attempt to fill the entire Universe?

It's because there is something 'containing' the gas – it's just not a physical barrier like the lid of a Tupperware. Instead, it's Earth's gravitational field, exerting a force on the atmosphere and keeping it in check around our planet.

I've already described how, as you get further away from Earth, the atmosphere starts to thin.[17] There isn't a hard and fast boundary beyond which the atmosphere 'no longer exists', there's just less and less of it as you get further out into space. This is why, if you ask the question, 'Where does the atmosphere stop and space begin?', you get different answers depending on who you ask – there isn't a universal definition and it's been the cause of a number of arguments in recent years. Although most people agree that the Kármán line, situated 100 km above sea

---

[17] The reason for this is perhaps not as obvious as it seems. Gas pressure pushes against the upper atmosphere, holding it above the lower atmosphere. The lower atmosphere can only support a thinner upper one. One good analogy is to think of a pile of sand, thick at the bottom and narrow at its peak. Try and invert it and gravity will very quickly return it to its starting position!

level, can be considered the 'start of space' as far as human activity is concerned, others define it as the mesopause (about 20 km closer). Wherever you define it doesn't really matter here – as long as we can all agree that space tourists and astronauts are two different categories of people, you'll get no complaints from me.

Aurorae
(For Real This Time!)

And now, a mere 8,000 words or so in, we get to the point: how and why do we get the beautiful lights at the poles of our Earth?

Let's surf on the solar wind as it leaves the Sun and journeys through space. Perhaps it was released as part of the Sun's normal day-to-day business, or perhaps some magnetic event flung it out there. Either way, it makes its way through the mostly empty interplanetary space, riding the IMF, until the incredibly hot plasma crashes into Earth's magnetosphere!

Now, as we know, much of it is diverted away from Earth around the magnetosheath. Those bits will continue their journey through space. Some of the plasma, however, doesn't fully pass Earth but instead gets caught up in its magnetosphere. Some of these charged particles get stuck (remember the Van Allen radiation belts?), but others ride the magnetic field lines to Earth's poles.

Well, not quite *directly* to the poles – it's a little more complex than that. If we look at a diagram of Earth's magnetosphere, we can see the sections of 'open' and 'closed' magnetic field lines that we discussed above, with the 'open' ones curving out into space. Rather than entering the magnetosphere at the centre of the 'open' section, though, the solar wind is directed along the boundaries between the open and closed lines. If we look at this from above, the areas where the solar wind enters Earth's atmosphere (and thus, the areas where we are most likely to see the aurorae) are shaped like an oval. I mentioned these regions earlier. They are the aptly (if not imaginatively) named auroral ovals. Depending on the IMF, the intensity of the solar wind and other 'space weather' conditions, the extent of these auroral ovals change. They're sometimes shaped like a thin ring, at other times they're spread out to cover wide swaths of the sub-Arctic and sub-Antarctic regions (see image 1). We can use geomagnetic data to 'draw' auroral ovals and there are many websites updated daily that aurora hunters use to see what the likelihood of spotting anything on a particular night is.

Earth's magnetic field is not completely static and unchanging. In fact, it's the complete opposite and changes constantly with the flow of particles from space. In addition, the magnetic fields inside the superhot plasma of the solar wind are constantly changing. While there are lots of different processes by which magnetic fields can rearrange themselves inside materials, the most important one for us is called **magnetic reconnection**. It does pretty much what it says on the tin: previously unconnected magnetic field lines connect to reconfigure the magnetic-field geometry inside that material.

These magnetic field events, particularly magnetic reconnection, are very important in determining the size and strength of aurorae. As part of these processes, some of the energy that was stored in the magnetic field is converted into other types of energy, including **kinetic energy**. Kinetic energy is crucial here, as it is related to the **velocity** (the speed of something in a particular direction) of the charged particles, which is known to affect the brightness and colours of the aurorae. In 2008, NASA's THEMIS satellites detected magnetic reconnection events at the same time as beautiful auroral

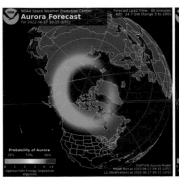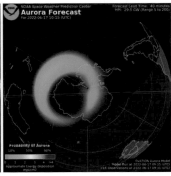

1. The areas where the solar wind enters Earth's atmosphere, and where aurorae are most likely to be seen, are called auroral ovals. The ovals can be mapped using geomagnetic data, as in these images from the NOAA Space Weather Prediction Center in the United States. These maps can help predict the likelihood of seeing aurorae at a certain time.

NOAA Space Weather Prediction Center

2. An example of diffuse aurorae, which can be spread across a large area of the sky. The unique perspective of this image, taken from the International Space Station, reveals that the red and green colours of the aurora are produced at different heights.
NASA/JSC/Scott Kelly

3. An auroral display captured in northern Russia. This image shows an example of an auroral arc with rays. The brightest areas are close to the horizon and rays of light spread upwards.

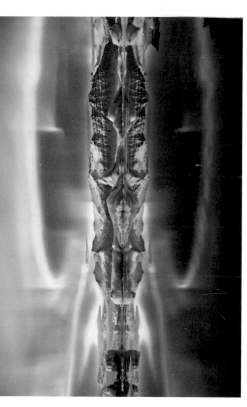

4. Another example of an auroral arc, captured in Iceland. Displays like this are also described as curtained aurorae. The reasons for the aurorae's distinctive shapes and movements are not yet fully understood.
© *James Woodend. National Maritime Museum, Greenwich, London*

5a and 5b. Two examples of corona aurorae, which occur in periods of high solar activity. Incredible views like these are only witnessed from directly beneath the aurorae. These photographers have skilfully managed to tease out some of the rarer blue and violet colours amidst the sea of green.
a. *National Maritime Museum, Greenwich, London. Courtesy of the artist.* b. *Crey – CC BY 2.0 (https://creativecommons. org/licenses/by/2.0/)*

6. In the Southern Hemisphere, people do not permanently live in areas inside the auroral oval, which means bright aurorae are rarely seen. Instead, a 'skyglow', like the one pictured here off the coast of Tasmania, is more common.

Dr Jenny Graff

7. The phenomenon nicknamed 'STEVE', or 'Strong Thermal Emission Velocity Enhancement'. STEVE was first spotted and photographed by aurora enthusiasts a few years ago. It is known to be distinct from 'classic' aurorae, but scientists are still working to understand more about it.
CDM Wild/Alamy Stock Photo

8. Aurorae have been identified on all the other planets in our Solar System, with the exception of Mercury. This image produced in 2007 shows pulsating, high-energy x-ray aurorae at the poles of Jupiter.

x-ray: NASA/CXC/SwRI/R.Gladstone et al.; Optical: NASA/ ESA/Hubble Heritage (AURA/STScI) displays were observed brightening the skies in the Northern Hemisphere! While these events aren't the only trigger for the aurorae, they do appear to initiate some of the more famous (and enthralling) 'dancing' behaviour.

I do have to introduce a significant caveat here: while there are a few different types of charged particles that make up the solar wind, for the purposes of the 'classic' aurorae,[18] we are only interested in the electrons. So, from now on, instead of saying 'solar wind', I'm just going to start saying electrons.

As electrons cascade through the atmosphere, they interact with each other and with the atoms and molecules that are already there. As we said earlier, collisions between electrons and atoms (or molecules) can excite those atoms (or molecules).

----
[18] Spoilers! I'll come to other types later on.

Depending on what the electrons hit (and where and how they hit it), atoms and molecules will be able to absorb energy from them and become excited. Once these newly excited particles transition to a lower energy level, either spontaneously or through subsequent collisions, light will be emitted at different energies, the results of which are the different colours of the aurorae.

In both oxygen and nitrogen, some of these transitions produce visible light in a range of different colours. There are numerous processes that give rise to these transitions and I'll outline some of them in the next few pages. This is by no means an exhaustive list, though – many require a little more physics and chemistry than the fundamentals I introduced earlier and others are still not well understood by experts!

Below you can see a table summarising the different wavelengths that we will talk about in this chapter, as well as their corresponding colour and source.

Particle	Colour of visible light	Wavelength (nanometres, nm)
oxygen (O)	green	557.7 nm
	red	630.0 nm
nitrogen (N_2)	blue-violet	391.4 nm
		427.8 nm
		470.8 nm
	red	665–685 nm

Oxygen

Let's start with the element that creates green, one of the most well-known auroral colours. There are a few processes that produce the beautiful greens we see in the sky, but one of the most dominant takes place thanks to oxygen. When an electron inside an oxygen atom decays from a very excited state down to a less excited (but still excited!) state, it emits a photon of 557.7 nanometres (nm) in wavelength in the process.[19] For reference, that's around 500 times smaller than your average dust mite. It's hard to get your head around the fact that just one type of atom out there, emitting one particular wavelength, is responsible for the bulk of the iconic green aurorae, but it's true! We'll talk a little more about why that is later on, but let's give a little history first.

When the wavelength of the green aurorae was first measured in 1868, it threw up quite a bit of confusion. The fact that different elements emitted light at different wavelengths was known and scientists were already eagerly measuring the absorption and emission spectra in an effort to explain

---

[19] Chemists will no doubt be crying at the fact that I am discussing single electrons here, as we're in fact talking about the oxygen atom transitioning between states (specifically moving between the two excited singlet states), but as a physicist who has spent too much time trying to make single electrons do my bidding, I think it's easier to visualise this way!

astronomical and atmospheric observations. It was also understood at the time that the atmosphere was mostly oxygen and nitrogen, so, naturally, it was assumed that auroral emissions would match their spectra. However, efforts to reproduce the green auroral emission with oxygen and nitrogen in a lab were not successful. For many, this null result didn't rule out the possibility that these elements were involved in the production of this colour light – they figured the experiment just needed to be improved. Others, though, went looking for different explanations – perhaps there was another element, never before seen, that created the aurorae! One of my personal favourite hypotheses was that krypton, a very heavy gas not found in any appreciable quantities in our atmosphere, was responsible.

In 1925, thanks to advances in technology, scientists were able to reproduce the green emission in the lab with oxygen. By that time, we'd just started to dip our toes into quantum mechanics and from the mathematical side of things there were some pretty strict rules about what states electrons could transition between. This gave physicists a false sense of security – they believed that, if we could calculate the energies at which different atomic transitions occur, then we could just look for them in nature!

Quantum mechanics, however, isn't nearly as well behaved as we would like it to be and there were a number of observed transitions that were simply 'not allowed' according to the rules at the time. You can't really argue with nature, but physicists tried anyway and gave what they'd seen the completely incorrect name of 'forbidden transitions'. While the name has stuck, we've since established that these transitions simply occur at a lower rate than others.

As you might have guessed by now, the transition inside oxygen atoms that would give us a 557.7-nm photon was one of these forbidden transitions. Specifically, it's the transition between the second-lowest excited state to the lowest excited state and once oxygen has been excited by electrons from the solar wind it happens pretty quickly – normally less than a second after excitation!

But the oxygen atom is still in an excited state, just a lower one than it was before. What about the transition from the lowest excited state to the ground state? Well, that emits light in the visible spectrum too – at 630.0 nm – but it is not nearly as bright as the 557.7-nm line. What does this mean visually? Well, light at 630 nm gives us red, as opposed to green. It's not as bright for many reasons, but one of them comes down to probabilities.

I've alluded to it already, but if an atom is left in an excited state for any length of time it will, eventually, decay to a lower energy state, emitting light in the process.

Similar to nuclear decay, you can't calculate exactly when this will happen, but you can give an estimate for how long on average it takes.

When it comes to spontaneous transitions, if a process takes a long time, then it's unlikely to occur without interruption and that is precisely the problem with this red transition. It takes a long time (relatively speaking) and so the oxygen atoms normally find some other way of getting rid of their excess energy, typically through collisions with other particles.

This means that down in the lower atmosphere the transition from the lowest energy state to ground state is very rare, but higher up it can and does occur more frequently. That's part of the reason why we don't see it as much. The dominance of green is also thanks to our eyes. They are much more sensitive to green light than red, meaning we will naturally see more of it. It serves to remember here that there is much less stuff (technical term) for the atom to crash into higher up in the atmosphere (it's extremely thin) and the atom is able to relax to a lower state,[20] emitting this red light in the process.

So, we've got oxygen's contribution: green light and red light. Green is emitted relatively high up in the atmosphere, between altitudes of around 100 to 150 km, but red is, for the most part, released even higher than that, as the red transition is suppressed at lower altitudes.

Nitrogen

Oxygen isn't the only element that gives us red light – nitrogen also has a part to play. In fact, nitrogen is responsible not just for some red aurorae, but also for the

---

[20] For the curious, this is through the oxygen atoms transitioning from the ^1D excited singlet state down to the ^3P orbital.

blue-violet colours that are both rare and absolutely stunning.

The colour produced by the nitrogen transitions depends on whether the atoms that make up the molecules are ionised (they have different numbers of protons and electrons). If the nitrogen atoms that make up the molecule have one fewer electron than proton then the resulting light will be in the blue-violet part of the spectrum, the colours most prized by aurora chasers both because of their rarity and beauty. There are three different wavelengths of light in the same part of the spectrum that can be emitted by transitions involving this particular type of molecule: 391.4 nm, 427.8 nm and 470.8 nm.[21] Here, rather than talking about a single transition, we're talking about multiple transitions from the same source. This is

---

[21] 'Airglow and Aurorae at Dome A, Antarctica', *Publications of the Astronomical Society of the Pacific*, 124(916), Sims *et al.*

because an oversimplification we made earlier is no longer valid – we can't restrict ourselves to a model with a single electron. Instead, we need to talk about the energy levels of several different electrons at once. Depending on which of those electrons are gaining or losing energy the wavelength of light emitted will change and we end up with many wavelengths close together. The blue nitrogen emissions typically occur quite low down in the atmosphere, at around 100 to 150 km from Earth's surface, but they can be seen higher up in altitude as well.

If we travel even higher in the atmosphere, though, we can see another colour emitted by nitrogen molecules. This time, we turn our gaze to the light emitted by neutral nitrogen atoms bound together to make N_2 and see red light, typically somewhere in the 665–85-nm range. Here we have a wider range of wavelengths than before because this transition is a result of the molecules losing their energy in collisions, **collisional *de*-excitation**.

So, there we have it. Nitrogen creates the prized blue-violet colour low in the atmosphere, as well as emitting in the red part of the spectrum on occasion. Oxygen moving from one very excited state to a slightly less excited state produces vibrant green and higher up we also get red as a consequence of it fully relaxing down to its ground state. Tiny, excited particles, making light shows to excite humans big and small.

Characterising aurorae

To this point, we've been thinking about aurorae from an atomic physics perspective, but for the sake of completeness (and because I want an excuse for beautiful pictures), we should turn our attention to the various different shapes and forms that they can take.

The exact reason for all these beautiful shapes is not very well understood, though we do have some hints. As mentioned earlier, the captivating, shimmering, dancing motion of the aurorae comes about thanks to magnetic reconnection events in Earth's magnetosphere. As the photons of light are emitted at various altitudes, the aurorae truly are three-dimensional, with both vertical and horizontal structure.

There are many ways to describe the shapes of the aurorae. Carl Størmer (1874–1957) was a pioneering Norwegian mathematician, aurora photographer and researcher responsible for much of our early understanding of how the charged particles of the solar wind could be moving through our magnetosphere and he attempted to create a classification system for aurorae in the 1930s. It was later refined for the publication of the *International Aurora Atlas* in the 1960s, but as time has gone on some of the categories described have been reformed, removed and redefined depending on whether they are used for photographic or scientific purposes (or both!). Some organisations discuss aurorae in terms of their form, others employ a two-part 'structure' and 'shape' definition, but there aren't really any fully agreed upon modern categories for different types of aurorae. There are some words, though, that are recognised as standard aurora descriptors.

Diffuse aurorae lack a structure and spread to fill a large section or even most of the sky. A photo of one such aurora taken from the International Space Station (ISS) (see image 2) offers a slightly different perspective to images captured from the ground, but it's still possible to distinguish between different shapes. After all, the ISS is orbiting Earth at around 400 km in altitude, so it's not all that high above where we find auroral activity.

What is abundantly clear from this perspective is that the colours are mostly produced at different heights – the red haze at the top comes from one of the oxygen transitions and contrasts with the green diffuse aurora lower down. There are also some ray structures visible towards the horizon line – the streaks of light are almost fence-like in their appearance.

The same shapes, this time viewed from Earth, are evident in an image of an auroral display in northern Russia's Yamalo-Nenets Autonomous Okrug (see image 3). Here we can also see an example of auroral arcs, with the brightest regions at the lower edges, marking out a snake-like squiggle across the sky with rays jetting upwards. You can see something similar in an image taken at Jökulsárlón, Vatnajökull National Park, Iceland (see image 4), though, in this case, instead of there being distinct upward rays we see more gentle ripples and folds, called 'curtain aurorae'.

When solar activity is low, you're more likely to get 'skyglow', a very faint haze across the horizon, or more diffuse aurorae. Conversely, the more high-energy electrons that come through the atmosphere, the more likely you are to get well-defined shapes and structures. What it is exactly that causes aurorae to create the shapes they do isn't yet understood, but the distinctive patterns and motions are likely down to a combination of atmospheric conditions and electromagnetic activity in Earth's magnetosphere.

Depending on where you view aurorae from (be that space or Earth), you will see different structures. The classic bands, rays, and curtains that we have looked at so far are the most common, but if you are positioned just so, directly underneath an auroral display, you'll be greeted by a fantastic sight overhead – a coronal aurora. They really are stunning, with beams that almost seem to come to a point directly above. These aurorae only occur when solar activity is very high and, of course, they only look like this when viewed from beneath (see image 5a) or very nearby (image 5b).

Finally, I couldn't leave this section without mentioning a stunning aurora picture taken from Tasmania, an island off the south coast of mainland Australia (image 6). As I've already said, aurorae are trickier to see in the Southern Hemisphere than in the Northern, but that doesn't mean they aren't there. In this image, we can see some rayed aurorae on the left-hand side, almost like searchlights pointing up, while across the rest of the landscape is a faint skyglow, reflecting off the water of the bay.

In addition to defining the structure of aurorae, we also categorise them by their brightness. While it's possible to use a photometer to get an exact reading, modern observers typically split them into four categories: IBC I, II, III, IV, where IBC stands for International Brightness Coefficient. IBC I corresponds to very faint aurorae where there's no real colour or likely any shape visible, while IBC IV is used to describe aurorae as bright as the Full Moon – bright enough to cast shadows! Aurorae of this type are very rare and most aurorae fall into categories IBC II and III – as bright as a thin, wispy cloud illuminated by moonlight at the faint end and like a fully moonlit, big, fluffy one at the brighter end of the scale.

There's another way to measure brightness too. Rather than considering the view from the ground, we can also take into account what satellites can measure from up in the sky. For this, rather than talking about the brightness of aurorae visually, we instead talk about the brightness in terms of the electron flux, which is the number of electrons passing by the satellite per second. This can be as 'low' as billions of electrons a second to anything up to 100 million billion – that's 10 with 16 extra zeros. This may seem excessive, but, when we're talking about things as tiny as electrons in something as large and spread out as our atmosphere, you need a lot of them if you want to be able to measure their interactions. That being said, you're unlikely to see electron-flux readings against beautiful professional photographs of aurorae – after all, we don't all have the means to check satellite readings to pinpoint activity in specific places at specific times!

You're familiar with my caveats at this point in the story, but here's another. This is not a definitive list of all the different colours and types of aurorae that we find on Earth – not even close. There are emissions in other parts of the electromagnetic spectrum, at energies not visible to our eyes, as well as other emissions in the visible spectrum that I haven't covered here. Oxygen can emit quite a few more wavelengths of red light than just the 630-nm line we discussed, and nitrogen has a whole host of other emissions that I haven't even dared touch on in this book, as well as some given off not by molecular, but by atomic nitrogen... If you're interested in the science behind those, you will need to go deep-diving elsewhere – probably into some pretty heavy chemistry first.

Events of high auroral activity

While our Sun is constantly emitting charged particles, visible aurorae occur when there is a huge amount, more than usual, of those particles being spewed out into space. This frequently coincides with events on the Sun's surface, such as coronal mass ejections or solar flares. By watching out for activity on the surface of the Sun, we can predict when we are likely to see lots of aurorae. Telescopes that watch the Sun can warn us of an incoming onslaught of particles, up to three days before they actually hit Earth! This is because the charged particles take between one and three days to reach us once they have been emitted from the surface of the Sun, so we see them being released before they reach us. This is really useful, not just for aurora chasers, but because when enormous numbers of particles hit our planet they can cause geomagnetic storms. These are distinct from 'normal' storms in that rather than being atmospheric disturbances that lead to rainy weather, geomagnetic storms are disturbances in Earth's magnetosphere. Aurorae are much brighter during these storms and are often visible at much lower latitudes than usual, because of the effects of the storms on the magnetic field of the Earth. Unfortunately, it's not all good news. Geomagnetic storms can affect the sensitive electronics on board the various satellites we have in orbit, potentially rendering them unusable. Even if there's no lasting damage, it gets much more difficult for data to be transferred between satellites during these storms, so there is the potential for massive communication delays.

Now this all sounds very doom and gloom, but it's not quite as dire as you might think. NASA, the European Space Agency (ESA), and various other government and non-governmental agencies are constantly monitoring the Sun and Earth, so they can mitigate the impacts of geomagnetic storms by activating shielding or even just turning off sensitive equipment, as well as a whole host of other methods too clever for me to understand (some of which are quite possibly classified). Thankfully for us, that means we can talk mostly about how they affect the pretty lights in the sky.

The Carrington Event is the largest geomagnetic storm on record. Precipitated by a colossal coronal mass ejection, when it struck Earth's magnetosphere late on 1 September 1859, it caused widespread blackouts and, on a lighter note, aurorae visible near the equator!

According to some reports, the aurorae observed in New York (a place that very rarely, if ever, sees aurorae) were bright enough to read the newspaper by! I can't even imagine that and I've spent a not-insignificant amount of time trying. If you've ever been outside on a really bright moonlit night, you may just about have been able to read some high-contrast text in moonlight (if you let your eyes adjust to the dark and then really strained them). Now imagine that, but replace the Moon with the aurorae. It's a sight I desperately wish I could have seen.

The solar flare that preceded the huge geomagnetic storm was witnessed independently by two astronomers, one of whom, Richard Carrington, was lucky enough to give his name to it.[22]

The Carrington Event was particularly impressive, but there have been plenty of others. At various times in the 19th and 20th centuries, auroral activity was reported as close to the equator as Cuba and northern parts of Australia! We should be wary of trusting every single reported incident. Historians of science are unsure whether every single account documents true aurorae, or if newspapers of the time were publishing something more sensationalist

---

[22] Really, though, they missed a trick. The other man was also called Richard (Hodgson) and I think they should have called it 'The Richard Event', after both men! They probably named it after Carrington as he was the Cambridge-educated professional astronomer, while Hodgson was more of an amateur observer, but still. Justice for Richard! The Richard Event it will be forever more in my mind.

and less factual. More recently, thanks to a particularly active Sun, aurorae were visible in the south of England for a number of days in October 2021 – much further south than you can normally see them!

While there are many gaps in our understanding of the processes that lead to prominent displays of aurorae, there are patterns evident in the timing of such events. One clear pattern is to do with the 11-year solar cycle we discussed earlier. Sunspot activity, solar flares, coronal mass ejections and therefore aurorae peak in the middle of this period, so if you want to increase your chances of seeing bright and beautiful aurorae it's worth keeping note of.

STEVE

I referred to the 'classic' aurorae above, but we cannot move on without a brief discussion of a newcomer to the scene, a strange phenomenon affectionately nicknamed 'STEVE', backronymed into 'Strong Thermal Emission Velocity Enhancement' (see image 7).[23] To begin with, a lot of the observations of STEVE were made not by professional scientists, but by people with a passion for aurorae.

There are many online forums and groups for aurora chasers to share tips and tricks for finding and photographing aurorae, as well as to show off their latest snaps. In 2017, Professor Eric Donovan from the Department of Physics and Astronomy at the University of Calgary came across a picture that he didn't recognise from the 'Alberta Aurora Chasers' Facebook group. Members of the group initially called it a 'proton arc', but

---

[23] Those who were around children in 2006 may recognise this as a reference to *Over the Hedge*, a film my siblings watched on repeat for an indeterminate, but extensive, period of time. Clearly, they were not the only ones enamoured by the name the animals in the story gave the mysterious new hedge: STEVE!

he knew that couldn't be right: for one, light produced by high-energy protons wouldn't be visible to the human eye or a camera lens. To their credit, some other members of their group had noticed this, and had given it the same 'Steve' – but that says even less about it! So, what was it?

Professor Donovan had access to data from all-sky surveys and specifically went looking through the data from Swarm. Swarm is a constellation of three satellites in low orbits near Earth's North Pole, studying the Earth's magnetic field and any changes that occur – in other words, exactly what Donovan needed! He obtained Swarm data from the night the image he'd seen was taken and was able to get another look. To Swarm, what the group was describing as a proton arc appeared as a wide ribbon of gas flowing through the atmosphere about 600 times faster than the air around it, with a temperature of about 3,000 °C(!).

Looking back through historical data for similar observations, he discovered that hot ribbons of gas had been recorded before, but scientists hadn't realised that they would produce a visual effect. In fact, it's possible that the plasma ribbons that we now refer to as STEVE had been observed by auroral researchers back in the early 1900s, but the observations were dismissed as being no different to 'standard' aurorae.

Without citizen scientists, STEVE would not be anywhere near as well documented or understood as it is today. There are tons of citizen-science projects, involving everything from classifying galaxy shapes to spotting new growth after bushfires have ravaged a region.[24]

If you find yourself somewhere where aurorae occur frequently, Aurorasaurus is

---

[24] Some of my personal favourites can be found on zooniverse.org, so check it out if you're looking for something to do!

a great citizen-science project for tracking auroral activity. Some people submit their own images with dates and locations that researchers then use alongside satellite data, while others simply register whether or not they could see aurorae in a particular place at a particular time. Aurorasaurus mines Twitter for mentions or pictures of aurorae and puts all of them on a map, a visual record of where the aurorae are visible at any given time.

Crowd-sourced data is a fantastic tool for researchers. In the past, if you wanted to compare visual and satellite data you would need to intentionally go out and collect it. You would align your plans with the movements of satellites, so that you could capture from the ground what they were recording from the sky, all the while hoping that the weather was on your side. Nowadays, thanks to projects like Aurorasaurus and all-sky surveys like Swarm, you can link tens, even hundreds, of pictures per night to satellite data. Professor Donovan himself said that findings like STEVE are some of the best things about the increasing inter-connectivity of the world and, for all the negatives, the fact that something completely new to science could be identified thanks to social media is a pretty amazing thing.

There hasn't yet been much scientific research into STEVE and what has been done hasn't revealed a whole lot about its nature. One thing is for sure, though: STEVE is definitely distinct from 'standard' aurorae! It seems to occur much higher up in the atmosphere and observations of the phenomenon show a lack of particle precipitation – that is, it doesn't appear to be caused by particles falling through the atmosphere. So, what is causing it? I'm afraid that's something for future aurorae books to figure out...

Aurorae on Other Planets

Aurorae occur because of an interaction between the Sun's emissions and our planet's magnetic field, so it is natural to ask, 'Well, what about the other planets in our Solar System? Do they have aurorae too?' The answer is yes (as you may well have guessed, considering I've dedicated an entire chapter to it)! In fact, the only planet in our Solar System that lacks aurorae is Mercury, though that's not because it has no magnetic field. It doesn't have any atmosphere, meaning that there's nothing for the solar wind to excite. The gas giants Jupiter, Saturn, Uranus and

Neptune do have strong magnetic fields and we have observed aurorae in their atmospheres – though they're not all identical to what we find here on Earth. The largest moon in our Solar System, Jupiter's Ganymede, has its own magnetic field too and, as it turns out, its own aurorae! Spacecraft have seen evidence of aurorae on Venus and Mars as well, despite the fact that neither planet has a strong, global magnetic field like Earth.

The aurorae on other planets are sadly not visible from Earth and even if we could hop over to Neptune for the night, or pay a visit to Ganymede, our eyes would not be able to see much, as the light emitted is, for the most part, not in the visible part of the electromagnetic spectrum. By studying these aurorae, we can learn more about the atmosphere and composition of their home planets, as well as seeing some pretty pictures along the way.

Let's concentrate on two particularly fun (in my opinion)[25] planets, with very different types of aurorae – Mars and Jupiter.

Mars

On Earth, the atmosphere and magnetic field are integral to the formation of aurorae. Without the large number of gas molecules in Earth's atmosphere with which to collide, charged particles from the Sun would not be able to produce any sort of light show and, without a magnetic field to guide them through the atmosphere, they wouldn't be present in such quantities! But that doesn't mean that planets without magnetic fields have no aurorae.

Mars, one of our nearest neighbours, is a planet with no appreciable magnetic field. It's also lost most of its atmosphere. From all that I've said so far, it's not

---

[25] And, since it's my book, I get to make the decisions!

exactly where you'd expect to find aurorae. And yet, in 2004, SPICAM (Spectroscopy for the Investigation of the Characteristics of the Atmosphere of Mars), an instrument deployed as part of the ESA's Mars Express mission, detected some local light emissions in the upper atmosphere of a particular region of Mars. It wasn't the sort of light that you could see with your eyes, but ultraviolet (UV) light. Now, if these light emissions had occurred during the day then this wouldn't have been anything special... but they were visible during the Martian night, suggesting that solar particles were somehow exciting the red planet's night-time side. The fact that light was being emitted in a very specific region gave scientists a hint when attempting to determine its origins.

Let's back up a bit: about 4.5 billion years ago, when Mars was first forming, it is thought that its core behaved much more like Earth's does – that is, rotating and generating a global magnetic field. For some reason or another, around 4 billion years ago this internal dynamo shut down, taking said field with it.[26] However, portions of the planet's crust are made of magnetic materials that have retained some magnetic field – quite amazing considering it's been billions of years! Previous exploratory missions that have mapped out the remnants of Mars's magnetic field revealed that most are found in the planet's southern hemisphere. This asymmetry suggests that the core stopped spinning while the planet was still cooling down, when the northern hemisphere was still hot enough to demagnetise but the southern hemisphere had cooled and 'frozen in' the magnetism.

---

[26] According to current (2022) knowledge, how or when this happened is very much up for debate, but it's a debate from which I choose to abstain!

ESA scientists overlaid their aurorae measurements onto a map of the local magnetic field on Mars and realised that the location in which they detected aurorae corresponded to a region of high magnetic-field intensity.

Over the next decade, SPICAM found further patches of auroral activity, which they named **discrete aurorae** due to their localised nature. As the researchers built up a better picture of Mars's magnetic field they discovered that the correlation between magnetism and the occurrence of aurorae wasn't quite as simple as it had seemed: a stronger magnetic field didn't always increase the probability of light emissions. In fact, in some of the most intense areas of magnetisation no discrete auroral activity was recorded whatsoever!

To explain this, we'll need to think back to our conversation on open and closed magnetic field lines. Recall that closed magnetic field lines are lines that connect a

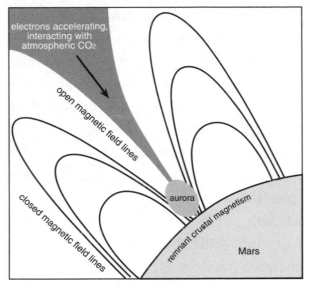

Figure 9: Portions of Mars's crust retain some magnetic field. Aurorae have been observed to occur on the boundaries between open and closed magnetic field lines.

north pole to a south pole, while open lines appear to originate from a north pole and continue out into space. It turns out that Mars' discrete aurorae were occurring on the boundaries between open and closed magnetic field lines (see Figure 9). In these areas, high-energy electrons from the Sun are accelerated thanks to transient electric fields, which are essentially short surges in electromagnetic charge. The electrons, upon entering the Martian atmosphere, collide with particles there, similar to the way in which aurorae are created here on Earth – though as the magnetic field is much weaker they are able to wander around the atmosphere much more freely than they do on our own planet. The Martian atmosphere, though, is mostly CO_2 and collisions between the charged electrons and CO_2 particles produce emissions not in the visible spectrum, but with UV wavelengths. It's a good thing SPICAM included a UV spectrometer!

When NASA's MAVEN (Mars Atmosphere and Volatile EvolutioN)[27] satellite entered the orbit of Mars in 2014 it

---

[27] Apparently a 'maven' is also a term for someone who is an expert in their field. Something tells me that someone searched the dictionary for words starting 'ma-' and was very pleased with what they found!

continued the hunt for aurorae. Within the first few months of its mission, MAVEN discovered that there was evidence of some faint aurorae on the night-time side of the planet, in areas without a large residual magnetic field. Due to their hazy and diffuse appearance, this type of aurora was named, you guessed it, 'diffuse'! These aurorae occur at about the same wavelength (UV) as the discrete aurorae, so must be the result of some sort of interaction between the solar wind and the Martian atmosphere. But if they weren't caused by some residual magnetism in the planet's crust, what generated them?

There may be no global Martian magnetic field for the solar wind to ride around the planet on, but in addition to those patches of crustal magnetism, there's the interplanetary magnetic field (IMF). Remember that the IMF is the part of the Sun's magnetic field that permeates throughout the Solar System.

It's thought to loop around Mars, allowing electrons and protons to surf the open magnetic field lines down into its lower atmosphere. Measurements from MAVEN indicated that the particle collisions precipitating the diffuse aurorae were occurring only around 60 km above the Martian surface – much lower than aurorae on any other planet that we know of!

So, that's two types of aurorae. Surely that's enough, right? Well, no. Just when we thought we'd found the key to aurorae on Mars, in 2018, signs of aurorae on the daytime side of the planet emerged too![28] If that wasn't enough, the auroral emissions were not at the same wavelength as those discovered previously, which means that we can be certain that different processes and particle interactions are

---

[28] To be fair, it was hypothesised in the early 2000s that Mars could have some sort of proton aurora.

involved. Fortunately, the wavelength of light being emitted was very well known: 121.6 nm.

Okay, so it's well known to physicists at least. You see, 121.6 nm is the **Lyman-α line**, a.k.a. the wavelength of light emitted when an electron orbiting around a hydrogen nucleus moves from the first excited state to the ground state. Bingo!

The mechanism behind this aurora is very, very different from any we have discussed previously. With the exception of STEVE, we've been discussing aurorae that are produced when electrons from the solar wind excite the atmospheric gases and cause them to emit light at a specific energy. Here, however, it's charged particles from the solar wind that are emitting the light, specifically the ones we've mostly neglected to this point: the protons.

'But hold on', you say. 'Didn't you just tell us that it was thanks to excited hydrogen atoms?' Well, dear reader, that's because the nucleus of a hydrogen atom is composed only of a proton. No neutrons, just a single proton. A neutral hydrogen atom is composed of an electron orbiting a proton and, if you remove that electron, you can call what's left either a proton or a positive hydrogen ion, H^+!

As protons on the solar wind reach the outer parts of the Martian atmosphere, they steal electrons from the hydrogen atoms found there, turning from H^+ (positive hydrogen ion, a single proton) to H (neutral hydrogen atom, single proton in the nucleus with a single electron 'orbiting' it). More often than not, these electrons are *not* in the ground state, so what we have are excited hydrogen atoms.[29] Now, as these newly neutral atoms move down through the atmosphere and the density of

---

[29] I really love talking about excited atoms. It makes me an excited atom!

the atmosphere increases, collisions start to occur. These can 'knock' electrons to lower energy levels, emitting a photon with an energy equal to the difference between the two energy levels in the process. This isn't the only way that the hydrogen atoms can de-excite, though. The electrons can also spontaneously drop to a lower energy level, once again emitting a photon! These processes result in a glut of emissions of a particular wavelength, which we call 'proton aurorae'.

Proton aurorae are not unique to Mars: they actually occur on Earth as well, but only in specific regions quite near the poles. Throw in the fact that they aren't visible and it's hardly surprising they're often sidelined in favour of the more showy, visible type. In fact, when STEVE was first discovered it was thought to be a type of proton aurora!

At first it seems quite amazing that a planet without an internal dynamo to produce a true global magnetic field is host to so many different types of aurorae. Really, though, some of these aurorae exist precisely because of the lacking magnetism. The fact that we know of these processes despite them occurring millions of miles away on a different planet, in light we cannot see, is testament to the work astronomers and engineers are doing to learn all we possibly can about the red planet.

Jupiter

If you were writing a list of things that were different on the planet Jupiter, you probably wouldn't really be thinking of aurorae – you'd be far too preoccupied by its lack of solid ground. But, different they are, and in some very interesting ways!

Jupiter's aurorae come in a range of different energies, with the highest-energy aurorae found on the planet having a much, *much* higher energy than those that we get on Earth. Forget about UV light – we're talking about x-rays! In 2007, NASA's Chandra X-ray Observatory, a space-based telescope that has been in use since 1999, took images of Jupiter at the same time as the *New Horizons* probe flew by the planet, allowing NASA to create composite images overlaying x-ray images from Chandra with visible light images from *New Horizons* (see image 8). The x-ray aurorae have been rendered as a purple haze above the poles. These x-rays are released in bursts with startling regularity. During a particular observation period ESA's XMM-Newton x-ray telescope, launched in 1999, observed a spike every 27 minutes over 26 hours!

It's a good thing I didn't write this book even six months earlier than I did, or I would have said that we weren't sure of the mechanism by which Jupiter was able to produce these pulsating, incredibly high-energy aurorae. Thankfully, an explanation was published in July 2021 by a team led by researchers at the Chinese Academy of Sciences, Beijing, and University College London.

Before I tell you the answer, though, I should probably outline why they were so hard to explain in the first place. Jupiter is bigger than the Earth in every way, so it shouldn't really be a surprise that the aurorae it produces are higher in energy, should it? Well, for one thing, the regularity of the pulses means that it's unlikely for the aurorae to be caused by processes like magnetic reconnection in Jupiter's magnetosphere, or transient events such as random coronal mass ejections or solar flares. On top of this, in order to emit x-rays you need really high-energy particles. Even with Jupiter's large magnetic field, models seemed to show that particles from the solar wind simply couldn't be accelerated enough to generate that much energy – they were just too light.

Instead, the researchers wondered if the solar wind was definitely providing the particles or if there might be something a little more local to Jupiter – perhaps something orbiting near the planet, inside its magnetosphere, spewing out relatively heavy particles on a fairly consistent basis...

If you haven't guessed already, don't worry, because I would never have thought of it either. It turns out that the origin of these heavier particles is one of Jupiter's moons, Io! Io is the most volcanically active body in the entire Solar System, thanks to its proximity to Jupiter's massive mass and magnetic field. Its volcanoes spray 'heavy' (to an astrophysicist) ions like sulphur and oxygen into the **plasma sheath** surrounding Jupiter, creating a **torus** of plasma encircling the planet. Juno, one of the man-made satellites that is in orbit around Jupiter,[30] has long observed this phenomenon. We can't take our explanation of Earth's aurorae and replace 'particles from the solar wind' with 'particles from Io' to describe the process though. If that were the case, we would expect the aurorae to occur in the same places as they do on Earth, within auroral ovals. In reality, however, they're found only at very high latitudes – basically *at* the poles.

Enter the aforementioned research groups. They arranged for simultaneous measurements to be taken by Juno and the XMM-Newton satellite to provide several different views of Jupiter from their respective positions. The XMM-Newton satellite is in orbit around the

---

[30] I absolutely love the naming of this satellite in particular. For those unfamiliar with Roman mythology, Jupiter is the king of the gods and his wife is Juno. Who better to send to check up on the mythological king than his wife?!

Earth and was able to observe any x-ray aurorae on Jupiter, while Juno was inside the plasma sheath itself, gathering data about its makeup and magnetic field.

Juno recorded fluctuations in the gas giant's magnetic field, observing **compression waves** travelling through it. These compression waves arose thanks to the solar wind crashing into and deforming Jupiter's magnetosphere. As this occurred, Juno also measured corresponding changes in the motion of the sulphur and oxygen ions responsible for the aurorae. The compression waves moving through the magnetosphere generated a special type of wave called electromagnetic ion cyclotron waves (EMIC), which were accclerating specifically those ions towards Jupiter's atmosphere.

When their results were combined, a beautiful picture was painted – the aurorae seen by XMM-Newton coincided with Juno seeing the compression and EMIC waves. There are still questions to be answered and further study to be done, but we do now have an explanation for how these regular x-ray pulsations might come about.

Afterword

As I noted very early on, it's quite easy to 'explain' aurorae: particles from the Sun are directed along the magnetic field lines of Earth, where they crash into our atmosphere, releasing light. Of course, that single sentence doesn't cover how the particles are created, why they go in the directions they do and what exactly they're crashing into to create all the different colours.

We have looked at all the stages of the aurorae, beginning with the subatomic particles of the solar wind, which ride the plasma through interplanetary space before finally being swept up by Earth's magnetosphere. From there, we followed these particles down through the maelstrom of our atmosphere, where they collided with various atmospheric particles, exciting them and producing a spectacular light show.

And, as it turns out, Earth isn't the only place where we find aurorae. By observing auroral activity on different planets, we can learn not just about the aurorae themselves, but also about the elements and molecules in the atmospheres of our astronomical neighbours. In the future, auroral studies could become part of atmospheric research of planets around stars other than our own Sun – though that's very far in the future.

While I have answered many questions in this book, I've certainly not answered every question you could have about aurorae. To be honest, I'm not sure that's possible, even if I were to write a book that was tens of thousands of pages long. We also do not know everything there is to know, certainly when it comes to aurorae on other planets.

Sometimes, people talk about science as 'ruining' things – they feel that knowing how something works destroys its magic. While an element of mystery is no doubt alluring, I don't think that gaining an understanding of something makes it any less beautiful. For me, knowing how the aurorae work only adds to the intrigue and fascination. I hope that knowing more about the forces and processes behind these incredible light displays makes them all the more beautiful to you too.

Appendix of Interesting but Not Necessarily Relevant Material

As I was writing this book, I found myself going off on the occasional tangent.[31] My zingy one-liners could be relegated to footnotes, but there are reams of information that, while fascinating to some, aren't necessarily pertinent here. My colleagues at Royal Observatory Greenwich suggested I write an 'appendix

---

[31] Okay, maybe they were more than occasional. Anyone who knows me wishes my tangents were only occasional.

of interesting but not necessarily relevant material' instead, which I immediately jumped at – it was that or delete a few thousand words.

I must warn you, I have not held back in the same way I may have earlier. In this appendix, for your reading pleasure, are a number of sidelines and tangents containing extraneous (but still compelling!) information relating to quantum physics, magnetism, history and what have you. I hope at least one person enjoys reading them!

History of our understanding of Earth's magnetic poles

Scientists haven't measured many changes in the location of the magnetic south pole, but over the past century alone the magnetic north pole has drifted from the northern parts of Canada to the Arctic Circle above Greenland, travelling over

2,500 km in the process! Over the past two decades, the movement seems to have happened at a faster rate, shifting between 50 and 60 km every year, or, in other words, 1 km a week. The British Geological Survey publishes official maps of Earth's magnetic field annually and they're used by many governmental and non-governmental organisations around the world. In 2020, however, they had to issue an early update due to the speed at which it was changing. While updating maps based on current data works very well, sudden changes have the potential to really mess with navigational systems and researchers are keen to search for ways in which they can predict the drift of the magnetic north pole and to look for other signs that may help forecast strange magnetic events in the future.

Data from observations of Earth's outer core are where researchers think answers could be found. A fundamental problem with our simple model of a spinning liquid core is that, in reality, it's not uniform. The ratio of nickel to iron is not consistent, neither is the density, the pressure or the temperature. Because of this, there may be areas of the core that move at different rates or even have different magnetic properties and over time this can influence the positioning of the poles, shifting them towards those stronger, more magnetised regions. Using satellites, researchers can observe changes in Earth's magnetic field and then model what sort of core changes might have taken place to cause them.

Some models suggest the presence of a sort of 'iron jet stream', moving from Alaska to Siberia, in Earth's outer core and while this explains some observed magnetic changes, it falls short of justifying the rapid acceleration that has been observed in recent years. It's too far north to adequately account for how the pole has wandered, so, while it's certainly affecting Earth's magnetic field, it doesn't quite hold the key to understanding why the magnetic north pole is moving so fast. Researchers at Leeds University suggest it's more likely that, over time, thanks to the non-uniformity of the flow of Earth's core, the magnetised region under Canada has weakened slightly, while the region underneath Siberia has strengthened and is therefore pulling the magnetic north pole towards it.

One thing almost all models have in common is that Earth's magnetic north pole isn't done moving yet – it's expected to continue its journey eastward towards Siberia, though whether it will continue to accelerate, stay at its current pace, or slow down again is up for debate.

What is also very interesting about this is comparing the magnetic north pole's change in latitude and longitude. The shift in its longitudinal position is happening much, much faster than its lateral movement (90° and about 14° respectively, over 100 years) and it's not a smooth change either. Most of that longitudinal shift has occurred since the early 1980s: 80° in the past 40 years! Observations suggest that between 1995 and 2005 its position was altered by over 4° in latitude, while in the next decade a movement of only around 1.5° was recorded. There was also a big longitudinal change during that same time period – over 18°!

Electron energy levels

One way of trying to visualise electron energy levels in atoms is by thinking of electrons bonding to atoms as being stuck down a well. The bottom of the well represents the ground-state energy. Any energy you gain will allow you to move further up the well towards freedom, with freedom meaning you are no longer in the well. Deeper underground, in a position of 'negative height' (i.e. -10 m above the ground is the same as 10 m below the ground), you will be closer to the ground-state energy, while the higher up the well you go the higher above the ground state you will be – until you reach 0 m, at which point you are free to leave.

The ground-state energy of an atom is the most negative. Successive higher energy levels are still negative but much closer to zero.

Another way of thinking about this concept is to consider the energy levels as a description of the amount of energy required to pull an electron away from the atom. An electron in the ground state, very close to the attractive nucleus, will require the most added energy to pull it away from the nucleus, so of course it's assigned the most negative number.

Consider an electron around a proton, otherwise known as a hydrogen atom. Its ground state energy is ~-13.6 eV, where 'eV' stands for 'electronvolt'.[32] The first excited state for hydrogen is around -3.4 eV. At first glance, it may look as though the ground state is at a higher energy as it has the bigger number but remember that we are talking about negative numbers. Unfortunately, like a staircase that would fail even a basic safety inspection, the steps between energy levels are not equal. If they were, the next step would result in an atom with positive energy: 6.8 eV.

What I said earlier, though, was that 0 eV represents a free electron, so how could the third energy level be 6.8 eV, a positive number?

Well, it's because I haven't given you the formula for calculating the energy levels of the hydrogen atom. Strictly speaking, we don't have one, but we

---

[32] For reference, it would take around 26,000 trillion trillion (2.6×10^{22}) eV to heat a litre of water by just 1 °C.

do have approximations that work very well (for hydrogen, at least, as it is the simplest case)! Despite having completely the wrong idea about how and why electrons could only exist in specific energy levels, in the first half of the 20th century, Danish physicist Niels Bohr (1885–1962) actually came up with a really good equation to calculate the energy levels of hydrogen:

$$E_n = -\frac{m_e e^4}{8h^2 \in_0^2} \times \frac{1}{n^2} = -13.6 \times \frac{1}{n^2}$$

This is a marvellously simple equation and it's impressive how well it works for hydrogen! Attempts were made to extend it to other elements, by multiplying the whole equation by the element's atomic number, Z, and, perhaps surprisingly, that works fairly well for some applications! But it doesn't hold for anything close to all atoms, and, even when it appears to work, it's only in certain circumstances.

For other particles, it really is nowhere near as easy and so, for the most part, we rely on approximations, measurements and the occasional tens of hours of supercomputing time. A lot of the time physicists will reduce problems to their simplest possible parameters: if you've ever heard a physicist joke about 'spherical cows in a vacuum', this is why – how much simpler can you get than spheres in a vacuum?!

Things get incredibly complicated from here, but there is a very general rule: the heavier the particle, the lower the ground-state energy. As you get heavier particles, the number of protons in their nuclei increases (meaning they are more positive), which will in turn attract the negative electron(s) even tighter.

If you are interested, there are a number of introductory books on quantum mechanics – just don't trust any of the ones that promise to let you see into the future or master the ability to use your quantum state to influence people. I can fairly confidently say that those won't be very useful.

Glossary

absorption spectrum (see **spectral line**)

atom – the smallest unit of ordinary matter.

atomic number – the number of protons inside the nucleus of an atom.

auroral oval – region around the poles where aurorae are typically most visible, so-named because they are oval in shape.

chromosphere – the layer of the Sun's atmosphere just above the photosphere, normally not visible but seen as a thin red line around the Sun during a solar eclipse.

collisional excitation – the addition of energy to an atom (or molecules) thanks to energy gained from collisions that results in it moving from a less excited state to a more excited state.

compression wave – type of wave that, when it moves through a medium, causes regions to compress and then extend outwards.

corona – the outermost layer of the Sun's atmosphere, visible during a solar eclipse

coronal mass ejection (CME) – an eruption of plasma into the corona of the Sun. Often accompanies a solar flare but they are distinct events.

diffuse aurorae – one of two types of aurorae seen on Mars. A diffuse aurora has no clear shape or distinct edge.

discrete aurora – one of two types of aurorae seen on Mars. This one has a clear shape and boundary.

electromagnetic (EM) spectrum – the range of all different types of light, from low-energy radio waves up to high-energy gamma radiation.

electromagnetism – one of the fundamental forces of nature, describing interactions between electrically charged and/or magnetic materials.

electromagnet – a magnet whose strength can be tuned based on the amount of electric current running through it.

electron – a subatomic particle, with a negative electric charge, found orbiting around the nucleus of an atom.

emission spectrum (see **spectral line**)

excited (state) – a state of a particle that is higher in energy than the ground state.

exopause – the boundary that marks the edge of Earth's atmosphere.

exosphere – the highest layer of Earth's atmosphere, from 800 km to approximately 3000 km in altitude.

ground state – lowest accessible energy level in an atom.

hydrogen fusion – the process by which hydrogen nuclei fuse together to form helium nuclei.

internal dynamo – the mechanism by which Earth's magnetic field is generated.

interplanetary magnetic field (IMF) – the magnetic field that is carried by the solar wind out into interplanetary space (i.e. between the planets).

ion – an atom with an unequal number of protons and electrons, i.e. a positively or negatively charged atom.

isotopes – atoms with the same number of protons, but a different number of neutrons in their nuclei.

Kelvin (K) – temperature scale often used in science, where 0 K is absolute zero, equivalent to -273.3 °C.

kinetic energy – the energy an object has due to its velocity.

Lyman-α line – spectral line energy emitted when the electron of a hydrogen atom transitions from higher energy levels down to the ground state.

magnetic reconnection – occurs only in very hot plasmas, when the orientation of a complex web of magnetic fields changes suddenly, resulting in magnetic field lines 'reconnecting'.

magnetosheath – boundary of the magnetosphere, where the magnetic field of the planet or star becomes much weaker and prone to disturbance.

magnetosphere – the area of space around a planet (or star) where its magnetic field is dominant.

mesopause – the boundary between the mesosphere and the thermosphere.

mesosphere – the third-lowest layer of Earth's atmosphere, from 50 km to 80–90 km in altitude.

molecule – a group of atoms bound together.

nucleus – the central, dense part of an atom, made up of neutrons and protons.

neutron – a subatomic particle, with neutral electric charge, found in the nucleus of an atom.

photon – a particle of light.

photosphere – one of the outer layers of the Sun's atmosphere, the one that is 'visible' to the human eye (though please never look directly at it!).

plasma sheath – layer of hot gas (plasma) surrounding Jupiter.

proton – a subatomic particle, with positive electric charge, found in the nucleus of an atom.

proton aurorae – aurorae created as high-energy protons falling through the atmosphere (of any planet) steal electrons from atmospheric hydrogen, emitting light in the process.

solar cycle (solar maximum, solar minimum) – 11-year cycle of solar activity, with the number of solar flares, CMEs, and sunspots peaking in the middle. Has been observed for over 200 years.

solar flare – intense, localised burst of light in the Sun's atmosphere. Often accompanies CMEs but they are distinct events.

spectral line(s) – energies of light that a particular element (or molecule) absorbs or emits. As there are many different lines for each element (corresponding to electrons jumping between different energy levels), each element has its own unique 'absorption spectrum' and 'emission spectrum'. These are the same spectrum; it is typically the absorption spectrum that is represented, however.

stratopause – the boundary between the stratosphere and the mesosphere.

stratosphere – the second lowest part of Earth's atmosphere, from 10–15 km up to around 50 km in altitude.

sunspot – an area on the Sun's surface (photosphere) that is cooler and looks visually darker. Not permanent, but appearing and disappearing over the course of days or months. More frequent during a solar maximum.

thermopause – the boundary between the thermosphere and the exosphere.

thermosphere – the second highest part of the Earth's atmosphere, from 80–90 km to approx. 800 km in altitude.

torus – doughnut shape.

tropopause – the boundary between the troposphere and the stratosphere.

troposphere – the lowest part of the Earth's atmosphere, up to around 10–15 km in altitude.

velocity – speed in a specific direction (e.g. 40 km/h north).